T0244862

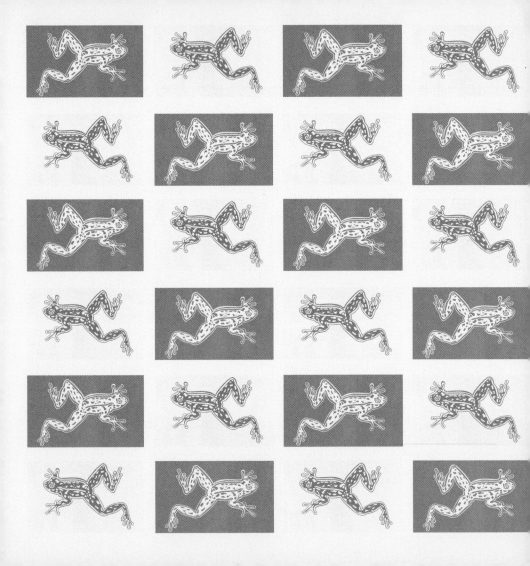

The Frog Book

by Jo Byrne

Series editor Jane Russ

Dedication

To Sean – my 'förg prunce' forever.

Contents

Introduction

I'd not given much thought to frogs at all until I met my partner, Sean. In late February, just a few months after we'd met, he suggested a walk through the woodlands near where he lived in Bedfordshire. 'Let's look for frogs!' he suggested happily. And there, in shallow ponds, we watched a large army of frogs in their mating 'embrace' (amplexus) – the air buzzing with sounds like little outboard motors as the amorous amphibians croaked away. I was enthralled! Sean, happy to have a 'frog pal', would also point out the toad crossing zones, as he regularly stops the car to help a hopster to safety off the road during migration season.

Just a few years later I moved to Bedfordshire myself and was delighted to discover a pond in the garden. It was bare and blank but had lots of fish. Over the course of the next few years, I filled and surrounded the pond with plants in an attempt to encourage as many bees and as much wildlife as possible. Before long, it was thriving with wildlife by day and by night. I was so thrilled, spotting first a toad and later a frog in the garden and then, the following spring, spawn in my pond. Sean drove over right away and we stared at the gloopy mush for ages. 'We've got frogs!' we said dreamily, in much the same way any couple would announce an imminent arrival of offspring.

Currently (summertime), the pond is full of the whole life cycle of frogs – from tadpoles with legs to impressively sized frogs peering at us from behind the pond plants, we've got everything going on. It is absolutely wonderful! A pond full of frogs is never still – there will always be a froglet swimming excitedly to a better sunning spot, or several 'teenagers' nonchalantly spread close to the surface, the occasional adult out on a hunting trip and the changeling tadpoles just... changing. Step outside the door and there's always movement, or something jumping out of your path.

Most excitingly, they've kept coming back! It's like Glastonbury here in late February! The spawn left behind is simply magical.

Magical. Transformational. Transitional. The frog. From tadpole to fully grown adult, this tenacious little creature undergoes a most fantastical metamorphosis.

Worryingly, Sean and I no longer find mating frogs or spawn on our woodland walks these days. All our regular spots have been bare for a few years now. Whilst Sean jokes that it's because they are all at my house, it's a stark indicator of the condition of the natural world. They are a critical part of our ecosystem and without their presence as both predators and prey, the fragile balance of nature risks a serious upset.

Did you know?

Frogs use their eyeballs to help swallow their food (see page 24).

All toads are frogs, but not all frogs are toads.

Between the 1940s and 1960s the African clawed frog (*Xenpus laevis*) was used by doctors as the most reliable pregnancy test.

Ancient Egyptians revered frogs so much they even had a frog-headed goddess, Heqet.

In this book, we are going to take a look at the frog and its magical transformation from tadpole to adult as well as looking at our native UK frog and toad species – of which we only have four – with a few non-native species now thriving too. We'll look at the worrying decline in frog species and numbers and just a few of the many threats to their survival. We shall also enjoy a romp through frogs in their various guises in myth and legend, as well as art and literature.

It is thought that frogs first evolved well over 100 million years ago. They've come a long way and seen incredible changes, morphing and adapting with those changes across landscapes, environments and ecological shifts and fluctuations. Kermit, perhaps the world's most famous frog puppet, once said, 'It ain't easy being green.' We can only hope they can continue to adapt, survive and indeed thrive.

Frog Physiology

Frog Physiology

Frogs are classified as amphibians – small, cold-bodied vertebrates (having a backbone).

They start life in water, using gills to breathe before undergoing a metamorphosis into air-breathing adults that can live both in water and on land. Amphibian comes from the Greek words *amphi*, meaning 'both', and *bios,* meaning 'life'. They belong to the order Anura, an ancient Greek word meaning 'without tail'. Frogs belong to the Ranidae sub order of Anurans and toads belong to the Bufonidae suborder of Anurans, giving rise to the oft quoted line 'all toads are frogs but not all frogs are toads'.

Here in the UK we have two native frog species, the common frog (*Rana temporaria*) and the pool frog (*Pelophylax lessonae*), and two native species of toads, the common toad (*Bufo bufo*) and the natterjack toad (*Epidalea calamita*). Peculiarly, you will not find the common frog in Jersey, of all places. Instead, Jersey has the agile frog (*Rana dalmatina*), which you won't find anywhere else in the UK! You'll also not find a common toad in Ireland.

All our native species of anurans can be found across Europe, with the natterjack toad being Ireland's only native species. In the UK, the rapidly declining number of natterjacks has earned them the rare full protection status, making it illegal to catch or even disturb one.

We have a few non-native species making more appearances in the UK, including the midwife toad (*Alytes obstetricians*), so named as the

male carries fertilised eggs wrapped around his back legs until they are ready to hatch. Confusingly, the midwife toad isn't a toad at all, but a frog, and the horned toad (genus *Phrynosoma*) is actually a lizard! Welcome to the crazy world of frogs and toads!

The Old English word for frog was *frox*, *froskr* in Old Norse, while Middle English words included *frosk*, *frok*, *frugge* and *froggen* (plural). In Scotland, frogs can be called puddocks, thought to descend from the Norse word for toad, *padda*. In North America tadpoles are sometimes called pollywogs. The collective nouns are an 'army' of frogs and a 'knot' of toads. People with sore, raspy throats will still use the phrase 'frog in my throat', and frogging fasteners for garments are still in decorative use in fashion today.

In the beginning

The first frog-like fossil is believed to date from around 180 million years ago, from the Jurassic Period of what is now South America and Antarctica, long before it became covered in ice. Part of the success of frogs in surviving throughout the millennia is their ability to adapt to the environment, resulting in different branching groups and over 8,000 frog species across the world, with more species being discovered regularly. They can be found throughout the world... with the exception of Antarctica.

Nice to meet you

Whilst belonging to the same order and starting their lives off in exactly the same way as tadpoles, frogs and toads have a few differentiating characteristics. Whilst they both breathe through their skins, shed their skins, eat the same foods and croak and ribbit, frogs prefer to be much closer to water than toads need to be. Frogs have a smooth, damp skin, whilst toad skin is dry and bumpy looking. A frog likes to leap and hop to get about and a toad prefers to walk. Toads possess poison glands that are not developed in frogs, but perhaps the most striking difference is their eyes, with the toad having a horizontal pupil and the frog having a round one. Their spawn is also a tell-tale indicator of whether you've found a frog or a toad. A frog will lay their spawn in big clumps whilst the toad lays spawn in strings. In addition to the different style of spawn, once hatched, toad tadpoles are poisonous whilst frog tadpoles are not. Hibernation requirements are different too, with frogs hibernating in mud at the bottom of ponds during winter whilst toads will bury themselves in leaves or damp soil.

Most frogs can change shades in their colouring to match their surroundings and enable better camouflage but are unable to change colour completely. This also helps with their body temperature, with some colours encouraging cooling and others warming. Like all amphibians, frogs will regularly shed their skin, but rather than hopping off after doing so, they will eat it! There is sense in this, as none of the nutritious protein in the skin goes to waste.

The slimy mucous covering of frogs helps keep their skin moist, enabling them to breathe, and they have several glands secreting moisture to keep the skin damp. In drier environments, evaporation means frogs quickly lose water through their permeable skin. At times like these, frogs 'drink' water by absorbing it through their skin from 'drinking patches' located on their bellies, then storing it in a bladder.

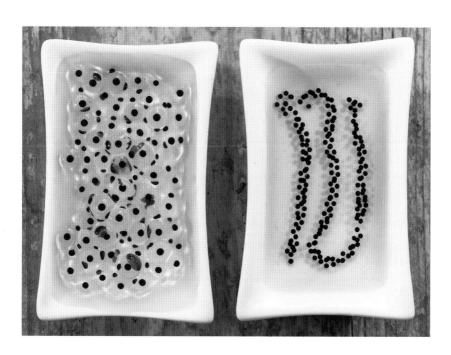

A frog will lay their spawn in big clumps
whilst the toad lays spawn in strings.

Frogs and toads aren't too fussy when it comes to food menus! Any unwary insect will make a tasty snack: ants, woodlice, spiders, beetles, moths. Frogs and toads also particularly love slugs, snails and earthworms. Whilst tadpoles are known to be cannibalistic, adult frogs and toads will rarely eat their own – only when food sources are scarce.

Here's looking at you

Frogs are canny, efficient predators but they also risk being predated by other larger animals, such as birds, foxes, badgers and hedgehogs, so they need to keep their wits about them. Their large, protruding eyes positioned on top of their heads are very useful in giving them a wide field of vision, perfect for keeping an 'eye' out for predators as well as spotting a quick snack. Frogs are myopic, or short-sighted, but have stereopsis, or binocular vision, allowing them to not only determine prey from foe but also providing them with a perception of depth in order to grab their prey with their tongue. However, in order to do this the frog must turn its entire body towards whatever has caught its attention, as frogs are unable to turn their heads. This becomes part of the fascinating ritual employed when a frog spots something. The frog will orientate its body towards something moving and focus both eyes on the item of interest. If the frog determines it is a foe, it will take defensive action, hopping off to hide or jumping into the water, but if the frog decides it's facing a tasty snack, it will stalk towards it, keeping its eyes focussed before snapping out its sticky tongue to grab its prey. Whilst frogs have teeth, these are used purely for holding onto the prey and, as the tongue of a frog is attached to the front of the mouth rather than the back, it is not used in the swallowing process. In order to force the whole prey down its throat, its eyes will close and the eyeballs will sink deep into the sockets, pushing the food down towards the stomach. The ritual is completed by the frog giving its mouth a wipe with a forelimb. Mmm, tasty!

Frogs have two transparent eyelids – top and bottom and a semi-transparent nictitating membrane – a third eyelid, so to speak – that enables them to keep their eyes open underwater. They also help keep the eyes moist when out of the water. As with mostly nocturnal vertebrates, frogs have a reflector system layer of tissue, *tapetum lucidum*, which bounces visible light back onto the retina, increasing the amount of light available and enabling better vision in the dark. In addition, it has recently been discovered that frogs can see in colour, even in what we would see as pitch-black. This ability is due to the rods in their eyes (the cells that work in dimmer light) having two different levels of sensitivity.

Hippity, hoppity!

Whilst they are famous for their jumping, in fact, not all frogs leap about. Toads prefer a sedate walk rather than a hop and will use their clawed hindlegs to dig themselves into muddy areas to hide or seek shelter.

The long back leg bones and foot bones in jumping frogs are fused, as are the front limb bones. This not only creates a stronger bone for leaping, but also helps cushion their landings. The muscles and tendons stretched across the back legs work in tandem with the tendons being stretched in anticipation of the jump, then spring back to help power the leg muscles in the jump forward – almost like a slingshot.

The fused tailbone of the frog – called the urostyle – works as a shock-absorber, allowing the power of the legs to propel the body forwards. During the leap, the frog also folds its front legs against its body and straightens itself out, allowing for a more streamlined and efficient use of its energy, before preparing to land. The webbing between their elongated toes helps to push the frog through the water as it swims.

Ribbit!

A frog's ear is located behind the eye and varies in size from species to species. It consists of a tympanum, or eardrum, and an inner ear. It is the tympanum you can often see on a frog and this acts to receive sounds and vibrate. These vibrations are processed by the brain and interpreted for meaning. As the tympanum acts like a drum, the size of the tympanum affects the male call – the bigger the drum, the bigger the noise.

The variety of frog calls is wonderfully wide and species-distinct. All members of the frog family are pretty vocal! Depending on the species, vocal sacs can be found either under the chin or at the side of the mouth. The frog inflates the vocal sac and squeezes air through the larynx in the throat. The spring chorus of frogs croaking will herald mating season and males calling for available females. It seems a female hears all she needs to know about a male from his call. Whilst it is more often than not the males that croak and call, some species of frogs have female callers. The female concave-eared torrent frog (*Odorrana tormota*) from China, emits an answer-call to male frogs to ensure she is easier to locate in the noisy waters where they live. Similarly, the male smooth guardian frog (*Limnonectes palavanensis*) in Borneo has a softer call and females respond loudly and more frequently. By contrast, the female African clawed frog (*Xenopus laevis*) emits a clicking noise to a male if she's not responsive.

Frogs can also emit an alarm call when bothered. In the case of the common frogs in my garden this is an unnerving high-pitched 'kissing' squeak when they find themselves the centre of attention of one of the cats. Fortunately, it is as alarming to the cats (and other predators) as it is to me, and the surprise element enables the frog to execute a quick getaway.

For toads, their protection from predators is found in the form of their parotid glands, bumps located behind their eyes. The milky bufotoxin produced in the glands is a mix of bufagin and serotonin, affecting the heart and vascular system, as well as bufotenine, which has a hallucinogenic effect. When feeling threatened, the toad will secrete the bufotoxin and effectively poison any would-be predators attempting to eat them. The toxicity levels vary amongst the different species of toads. The Australian cane toad (*Rhinella marina*), for example, is extremely poisonous but, thankfully, here in the UK our *Bufo bufo* is much less so. It is still wise to be cautious, however! Sometimes dogs will try their luck and this can result in foaming at the mouth and vomiting. If this happens, it is important to get the dog to the vet as soon as possible.

Australian cane toad

Mating

During spring, our frogs and toads migrate their way back to the waters where they were spawned. Whilst frogs generally don't stray too far from their water source, this is a very dangerous time for toads as they attempt to navigate traffic, plodding along roads determined to make it 'home' for breeding season, and they can travel a considerable distance. There are suggestions that frogs and toads use a variety of senses to find their way back, including being drawn to certain smells or suitable plant material that will feed the tadpoles as well as sounds of males already in place and beginning to call. Furthermore, it is thought that they are capable of recognising landmarks, which is why frogs and toads will often return to gardens even after a pond or water feature has been filled in. They will

usually make their way to a nearby alternative – if available – should their original 'home' no longer be welcoming.

In the warmer parts of the UK, mating can start as early as January, while in the colder parts, like Scotland, it might be as late as April. Common frogs normally have a week or two head start on common toads. However, natterjack toads

wait until April and pool frogs until around May. Around my pond in the South East Midlands of England, we normally expect our flurry of visitors 'after the first warm rain' in late March, and I've seen spawn as early as February one year, which, sadly, didn't make it through a subsequent cold snap. In 2022, it was clear the ferocious storms Dudley, Eunice and even Franklin had absolutely no affect on our amorous anurans, as we found huge clumps of spawn across the pond during that crazy February – it proved a very successful year for tadpoles.

The males will normally be more numerous than the females and will arrive at the breeding ponds first and begin their chorus of croaking, calling for available females. The deeper the call, the bigger the male, although, thankfully, the female is normally larger than the male (at least during breeding season). Once the male finds an available female, he's determined to hang onto her! Amplexus – Latin for embraced, clasped or grasped – is when the male climbs onto the back of the female and holds on, locking his arms underneath her forearms. In male frogs, this embrace is helped by extra sticky 'nuptial pads' on the thumbs. In this position, the male is then in the best possible place to be ready to release sperm onto the eggs once released by the female.

Frogs and toads are not normally territorial or territorially aggressive – until mating season. With fewer females than males, other males might try to kick off or pull males from the female they are locked onto. Worse, more males climb aboard and, in these instances, females can drown.

As the first life stage of frogs and toads is aquatic, the eggs – called spawn – are laid in water. In frogs, spawn is deposited in large clumps, and a common frog female can lay between 1,000 and 2,000 eggs. Toads lay them in long strings a little deeper in the water and wrap them around submerged vegetation. Common toad females lay between 3,000 and 6,000 eggs as a double chain whilst natterjack toads lay their spawn as a single chain.

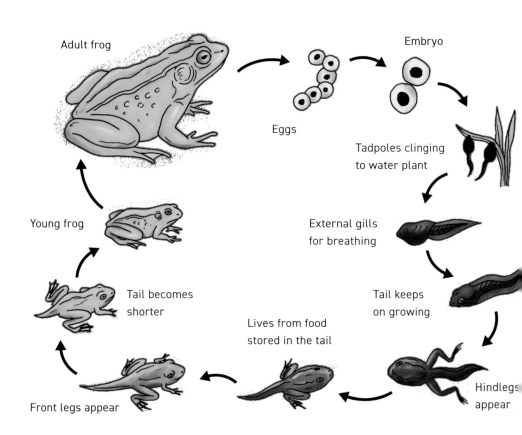

Adult frog

Eggs

Embryo

Tadpoles clinging
to water plant

External gills
for breathing

Tail keeps
on growing

Hindlegs
appear

Lives from food
stored in the tail

Front legs appear

Tail becomes
shorter

Young frog

The magical metamorphosis

The spawn is laid and the extraordinary journey begins, though not without incredible dangers for the young anurans. On average, only 2% of eggs make it to tadpole stage. From the moment the eggs are deposited, they are under threat from predation. In those eggs that have been laid somewhere safe, the little black dot in its ball of jelly is transforming at a rapid pace and using the jelly glycoprotein as sustenance.

In just a few days, with decent weather and temperatures, it is developing a tail, head and gills and beginning to wriggle about. A week or so later, the little tadpoles move out of their bubble but remain on top of the jelly for a few more days. The jelly starts to sink into the water and once set free the little tadpoles face an even more vulnerable stage due to predators – including each other!

At first, the tiny tadpoles are herbivorous, scraping off algae and plant matter with beak-like mouths, but as they develop, and they and their intestines shorten, their diet becomes more varied. They will eat dead animals in the water if the opportunity presents, but also other tadpoles! In ponds or water features that have a glut of tadpoles, larger ones will happily kill and eat their smaller siblings.

Toad tadpoles are poisonous to predators, already developing the toxins they'll use in defence when they are older, whilst frog tadpoles are not. In frogs you are more likely to see tadpoles swimming together in an effort to minimise being picked out by a predator, whereas toad tadpoles will happily strike out on their own.

Over the weeks, the tadpole morphs into what will be their land-self: their back legs develop and overall the tadpole's length shortens. Front legs appear, the tail is reabsorbed back into the body and, importantly, the tadpoles change from water-breathing creatures with gills to air-breathing froglets with permeable skin and lungs.

The whole process takes between 10 to 16 weeks for common frogs and common toads, depending on the weather conditions. That said, because natterjack toads depend on very temporary water pools, their transition from egg to toadlet can take place in as little as six to eight weeks.

The mortality rate from egg to adult is ridiculously high, hence the large number of eggs laid by the female. Of the 2% of eggs to reach tadpole stage, less than 1% make it to froglet stage. The sparse number that make it out of the pond must avoid being predated on by crows, magpies, blackbirds, herons, hedgehogs, badgers, grass snakes, otters and even succumbing to dehydration if caught out too far from their water source.

Less than 1% of froglets make it to adulthood. If they make it that far and can avoid their everyday predators and disease, a common frog can live between 5 and 10 years in the wild and a common toad between 10 and 12 years, but it is an very precarious journey!

Take a deep breath

Young tadpoles start off with external gills to extract oxygen from the water. After a couple of weeks the gills become covered with skin and the tadpoles continue to breathe underwater using internal gills.

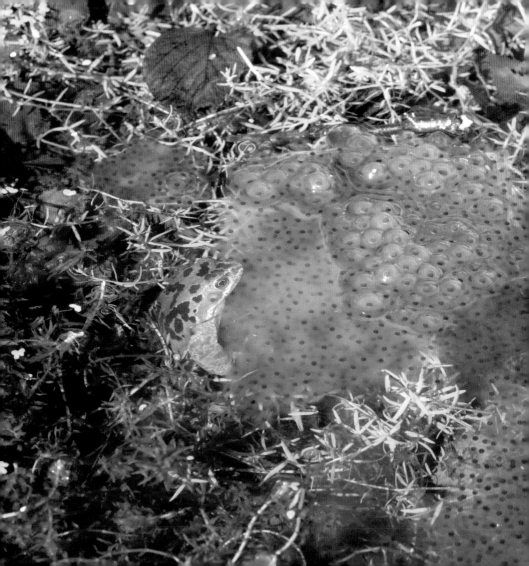

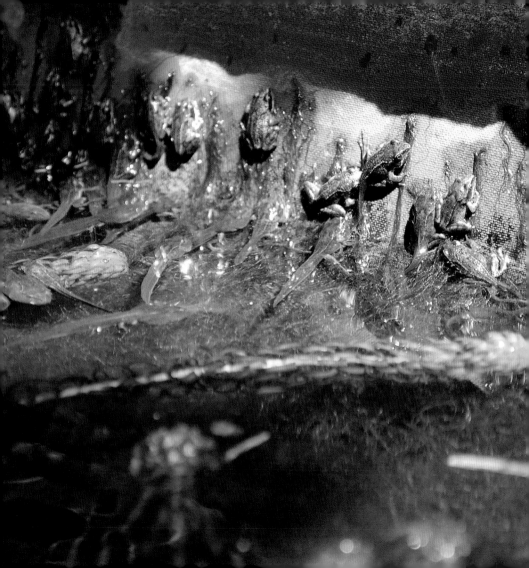

Around the same time as the legs begin developing, the lungs develop, and you will often see tadpoles breaking the surface of the water and gulping air. Once they've reached froglet stage, the gills will have been absorbed back into the body, just like the tail, and the young frog will rely solely on their skin and lungs for breathing.

As frogs do not possess ribs or a diaphragm to help with the mechanism of breathing, the frog's mouth remains closed but it lowers the bottom part of its mouth (or buccal cavity), opens its nostrils and allows air in. Then the frog closes its nostrils, contracts the buccal cavity and forces the air down the throat into the lungs. To breathe out, the buccal cavity is lowered, the carbon dioxide drawn from the lungs, the nostrils opened and the cavity contracted, pushing the air out.

A frog also breathes though its skin, which is thin and permeable and allows the easy diffusion of respiratory gases. When underwater, frogs use only their skin for breathing. As such, frogs need oxygenated water, particularly when submerged in the mud, slumbering through winter. Without oxygen in the water, the frog will drown.

Time for a nap

Whilst people often suggest that frogs and toads hibernate during winter, the correct word is 'brumate' for our cold-bodied amphibians. During this time, their metabolism will slow down and they go into a state of dormancy during the coldest times of the year. Common frogs will bury themselves in mud at the bottom of ponds, while pool frogs and common toads cover themselves with leaves or soil and natterjack toads bury themselves in sand. Compared to the hibernation of warm-bodied animals, brumation does not last as long. The other difference between hibernation and brumation is that hibernating animals tend to eat a huge amount before the cold weather sets in, allowing them to feed off their fat stores during dormancy. Frogs and toads, however, eat less before their dormant time but will nip out and forage from time to time on warmer days, looking for something to nibble and something to drink. I've had first-hand experience of this! Early one mild winter's evening in December, I was amazed to find a huge toad sitting on the path. He looked at me for a bit and then began lumbering up the leafy undergrowth on the side of the road, heading for the woods. I looked for any obvious injuries or problem and could see none. Then, having made sure he/she was safely heading away from – rather than towards – the busy road nearby, I wished them a merry Christmas and carried on walking.

UK Native Frogs and Toads

UK Native Frogs and Toads

Common frog

The common frog (*Rana temporaria*), as the name indicates, is quite common around the UK, found in slow or still ponds, rivers and lakes. They have been known to monopolise

tiny urban water features and spawn in any available watery receptacle. Common frog adults are around 6-9cm in body length (not including their long legs) and, whilst males and females are difficult to tell apart, females tend to be bigger, particularly during breeding season. The males will have a hint of blue on their white throats at this time. Overall, they have rounded snouts, with their vocal sacs underneath their mouths. Black splotches on their skin behind their eyes look almost mask-like and, whilst they are normally brownish with black mottling across the body and black banding around the legs, they can vary in colouring and even have red, grey, olive or orange speckles. The spotting of the skin is as individual to them as a fingerprint is to a human.

Pool frog

The pool frog (*Pelophylax lessonae*) is a little smaller than the common frog, with a pointier snout and vocal sacs at the side of the mouth rather than underneath. They are distinguishable by a yellow or green stripe along the middle of the back, although it is currently unlikely you will see one in the wild for some time to come. The last 'true' pool frog, Lucky, died in captivity in 1999 after being captured from a pond in Thetford, Norfolk, effectively meaning the pool frog was extinct in the UK. Thankfully, conservationists found Swedish pool frogs were a very close genetic match for our pool frogs and a successful reintroduction programme in a managed area of East Anglia means there is some hope that pool frog numbers might increase and become more widespread in the future.

Common toad

The common toad (*Bufo bufo*) is larger than the common frog at around 8cm from nose to tail, but they are definitely a bit shyer than the common frog, preferring forests and moorlands or quiet gardens, hiding underneath old logs and big stones. Toads don't need to be as close to water as frogs and can find their hiding spots a good distance from watering holes. Common toads have bright orange eyes with a horizontal pupil and prefer to walk calmly rather than hop. Their dry, bumpy skin will show a variety of colours ranging from grey and brown to black and yellow, occasionally even red. The vocal sac is under the chin, with the poison glands, or paratoid glands, evident as lumps behind their eyes. Toads prefer to return to their original places of birth in the breeding season and their annual spring migrations see them trek all the way back to their 'home' ponds. Thankfully, countrywide Toad Patrols undertaken by volunteers help the dusty night-time travellers plod on across roads on their determined journeys.

Natterjack toad

The natterjack toad (*Epidalea calamita*) is a little smaller than the common toad but similar in terms of its bumpy skin and raised paratoid glands behind the eyes. Its colouring is a more standard-patterned brown and white mottling but with a very distinctive yellow stripe running down its back from head to tail. The natterjack's wonderful eye is either gold or green with a black marbling effect, as well as the horizontal pupil characteristic in toads. Much like the pool frog, the continued presence of the natterjack toad in the UK is due largely to the efforts of conservationists and protected sites all around the UK. These sites are mostly coastal, as the toads prefer to live in sand dunes, seeking out small pools for mating and laying spawn.

The name 'natterjack' literally means 'chattering toad', from an old word for a male or small animal (jack) and natter, or to chatter. The natterjack certainly knows how to 'chatter' with a very loud, raspy call that can be heard up to 2km away – as such, other nicknames for them are the 'Birkdale nightingale' and 'Bootle organ' from those respective areas in the north-west of England.

Marsh Frog.

Our Non-native Frogs and Toads

Our Non-native Frogs and Toads

Unfortunately, the habit of keeping frogs and toads as pets has meant that, for years, folk would thoughtlessly discard them into the wild – or accidentally lose them! Thankfully, the practice of releasing non-native species is now illegal in the UK, but we've still a few species that, in true anuran style, have simply adapted to their new home.

Midwife toad

The midwife toad (*Alytes obstetricans*) is naturally found in Europe and northwestern Africa but accidentally arrived as a stowaway in Bedfordshire in 1904, where it has settled and thrived and now even been found in Yorkshire and Devon, though, thankfully, their really small size means they don't particularly represent a threat to other species. They are recognisable by their high-pitched squeak call and are mistaken for toads due to having the warty paratoid bumps behind their eyes. Their midwife name is derived from the behaviour of the male, who carries the fertilised eggs wrapped around his back legs to keep them safe until they are ready to hatch, at which point he then returns them to the water. Midwife toad tadpoles are unusually large (longer than the adults) but, like all other frog tadpoles, are not poisonous.

The green frogs

The marsh frog (*Pelophylax ridibundus*) and the edible frog (*Pelophylax esculentus*) are known as green frogs, a name also associated with our native pool frog. The larger marsh frog first settled around the Romney Marsh area but has successfully spread across Kent and even into Sussex. The smaller edible frog is indeed edible and known

Midwife toad.

Marsh frog.

American bullfrog.

African clawed frogs.

to be making its home across Surrey and into Sussex. The green frogs have quite noisy mating calls and usually breed a little later than other frogs, even as late as May.

American bullfrog

The most troublesome non-native frog is the American bullfrog (*Lithobates catesbeianus*). With a distinctive bellowing call to match its very big size, adults can measure between 15-20cm from nose to tail and weigh up to 500g. The bullfrog will happily gobble up anything smaller than itself – including smaller frogs – and wild colonies are of grave concern to conservationists.

African clawed frog

The African clawed frog (*Xenopus laevis*) is fully aquatic and rarely leaves the water. It has been spotted around Lincolnshire, Wales and on the Isle of Wight and is generally believed to have escaped – or been deliberately released – from laboratories, where they were used for pregnancy testing from the 1940s to the 1960s. It was discovered that urine from a pregnant woman injected into the back of the frog would cause the frog to lay eggs within a matter of hours, a much speedier result than the previously used animals (mice and rabbits), plus the hardy frog could be used repeatedly. Sadly, experimentation on the African clawed frog continues to this day, including them being the first cloned vertebrate. Unfortunately, as voracious feeders, the African clawed frog is considered a menace to native species.

Poison dart frog (*Ranitomeya amazonica*).

Frogs Around the World

Frogs Around the World

Poison dart frogs

There are around 250 different species of poison dart frogs found around Central and South America. Many are strikingly and beautifully colourful, but this, in fact, is meant to act as a warning and deterrent to potential predators, a trait known as aposematism. The frog's toxin comes from their diet of termites and ants. Local Native American tribes run their darts over the skin of these frogs and then use the poisoned dart in their blowguns for hunting.

Golden poison frog

In Colombia in south America resides the golden poison frog, whose nickname is 'the terrible one' (*Phyllobates terribilis*). Its beautiful golden skin is covered in a poison so potent that a person could die from a single touch and which can last for up to a year on hunting darts. It can also be used (in VERY small quantities) to reduce chronic pain.

Strawberry poison dart frog

Dart frogs make great parents! The female strawberry poison dart frog (*Oophaga pumilio*) from Panama normally remains on the ground and will lay her eggs among fallen leaves. Once the egg hatches, however, the tiny tadpole will climb onto the mother's back and the female frog, despite not being well suited for climbing at all, scales trees in search of tiny puddles of water between the spiky leaves of bromeliad plants growing up there. Once a suitable spot has been found, she will let the tadpole slide off into the puddle, repeating this for all of her offspring (between four and six at a time). By releasing them into separate trees and puddles,

Golden poison frog.

Strawberry poison dart frog.

she increases the chances of some of them surviving should one or two be discovered and predated. Fascinatingly, the mother will return periodically to each of the watery hiding spots and lay unfertilised eggs in the water for the tadpoles to feed on. Her recall for where each and every one of her tadpoles are is faultless. The eggs she feeds her tadpoles also contain the alkaloid toxin the adult frogs have, helping provide the offspring with a degree of protection against being eaten by predators. The male strawberry dart frog co-parents in this whole magical endeavour by defending the nests where he can and keeping the water levels of the puddles topped up – that is, he empties his bladder into them!

There are some other quite wonderful frogs around the world!

Surinam toad

Like the African clawed frog, the Surinam toad (*Pipa pipa*) is fully aquatic but the male can still make himself heard to available females even underwater. His call is more of a click than a croak as he 'snaps' his hyoid bone in his throat. During amplexus, the male and female will spin and flip repeatedly in the water until the eggs are finally released and fertilised. The male will then embed the eggs into the skin of the female, across her back.

The eggs sink into the skin and form pockets where the embryos are safe. They develop through the tadpole phase and emerge as young toads 12 to 20 weeks later. After the young toads have disappeared off to live their solitary lives, the adult female will shed her skin and be ready for the next season.

Surinam toads are also known as the star-fingered toad due to the shape of the protuberance at the end of their 'fingers' – these lobes are used for feeling out prey in the murky rainforest waters.

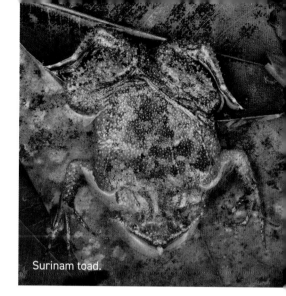

Surinam toad.

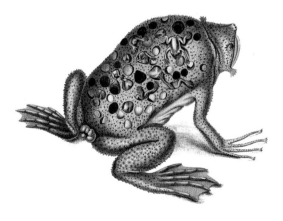

Wood frog

The wood frog (*Lithobates sylvaticus* or *Rana sylvatica*) is found across the northern parts of the American continent including Canada and Alaska. It is currently fascinating cryobiologists for its ability to withstand freezing temperatures for extended periods – several months, in fact! A combination of urea and glucose is produced, creating a 'natural anti-freeze' to protect its vital organs and allowing it to hibernate in icy conditions, reducing the body temperature as well as the heart, breathing and metabolic rates.

Desert rain frog

Possibly one of the cutest frogs, the desert rain frog (*Breviceps macrops*) is also the only frog which misses the tadpole stage, instead developing directly from eggs laid in their damp, sandy burrows into fully formed frogs. They emit a particularly high-pitched warning call which sounds not unlike a squeaky toy!

Dancing frogs

Frogs of the Micrixalidae family of the mountainous areas of western India are known as 'the dancing frogs'. It is believed that, due to the noisy waterfalls in which they like to live, they've developed a better technique for attracting mates – 'foot flagging'. In addition to having a very prominent white vocal sac to catch an available female's eye, the male frog starts by tapping his hind leg before stretching it out and wiggling his toes! He'll then repeat the dance with the other leg.

Hairy frogs

Finally, the rather incredible adult male hairy frog (*Trichobatrachus robustus*) is found in central Africa and develops hair on its flanks and thighs during breeding season. It is thought the hair assists with breathing as the male remains in the water with the eggs for some time after fertilisation. It is also known as the 'Wolverine frog' and, like the Marvel character, the frog effectively breaks its toes and forces the sharpened bones through the skin as a defence mechanism.

Common toad

Threats and Dangers
to Frogs

Threats and Dangers to Frogs

As part of the food chain, frogs and toads not only have a long list of insects and small creatures they like to nibble on but likewise find themselves on the menu for many other creatures.

They are not even safe in the water! Newts, fish and large diving beetles happily feed on tadpoles. Dragonfly and damselfly larvae are carnivorous and voracious feeders that also feast on tadpoles. If the tadpoles live through all those dangers and make it onto land as froglets, they then must dodge being eaten by all the various animals that will readily help themselves to the amphibians: hedgehogs, crows, blackbirds, badgers, foxes, otters. Otters avoid toad poison by peeling the skin off, but grass snakes are immune to it. Whilst frogs will make a dash for it by hopping away from grass snakes, toads are known to fill themselves with air and push themselves up onto their tippy-toes to try to appear as big as possible in the hope of convincing the snake that it is too big for it to consume.

If the constant risk of being eaten wasn't enough to contend with, there's the risk of parasites, fungal infections and viruses to worry about. Whilst, like most creatures, frogs and toads will be host to and affected by a variety of parasites and infections, there are a few here in the UK that are particularly harmful to our native UK species.

For our UK common toad, one of their biggest enemies is the toad fly (*Lucilia bufonivora*). The insect will lay eggs on the toad around the nasal passages

This lucky frog managed to escape this
tricky situation!

Example of toad fly infection.

A frog affected by chytrid.

and, once hatched, the larvae climb into the nasal passages and start to feed on the amphibian from the inside, which invariably kills the hapless toad. The larvae then bury themselves in the soil to pupate and ready themselves for the next season. Once the toad fly larvae have infected a toad, there's no cure and it will die. As such, this is a particularly worrying parasite for conservationists – and toads!

Batrachochytrium dendrobatidis (often called chytrid) is a hugely worrying and extremely deadly fungus thought to be wiping out amphibian populations around the world, infecting frogs and toads and, recently, a new strain has been observed killing off salamanders. The fungus grows across and into the skin of the frog, interfering with the gaseous exchange needed to breathe. It also upsets the water, salt balance and excretory systems of the amphibian, eventually causing heart failure. In humans, it would be the equivalent of attacking the lungs, kidneys and skin in one fell swoop. This fungus is thought to be devastating populations worldwide and there is currently no treatment for it. Once the fungus has infected a pond or waterway, it will kill off all adults and young. The tadpoles will survive the fungus only until they metamorphose, at which point the fungus burrows into their skin and kills them as well.

A particular threat to the common frog around the south-east of England is the ranavirus. Whilst difficult to detect prior to death – other than possibly being able to spot lesions or red blotching on particularly swollen frogs – the main clue of an infection is simply a sudden mass die off of frogs and, worryingly, the virus can spread very quickly. If your pond falls victim to a sudden die off of frogs, please report it immediately to Garden Wildlife Health via their online reporting tool: www.gardenwildlifehealth.org.

There is some evidence that our UK natives are building up resistance to the virus, but it remains deadly and is known to be able to survive for some time in water, as well as, it is suspected, in the soil.

Then there is the pervasive threat of humans. From destroying natural habitats, harvesting them for food and even still using them in scientific studies, amphibians have a lot to be worried about when it comes to us.

There is plenty of archaeological evidence that frogs were eaten across Britain and Europe as far back as 7,000 BC, and this trend continues, from the famous French delicacy of frog legs to frog soup served across Asia. Whilst Indonesia has traditionally been the world's largest exporter of frog meat, New Zealand reported a staggering export value of nearly $204 million in 2021 and a market share of just over 33%, with Europe being the largest

importer of frog legs. The concern for most conservationists, particularly with exports from Indonesia, is that much of the several tonnes of frogs exported each year are wild harvested rather than farmed. Apart from the devastating effects on the frog population, there's the delicate ecosystem to consider: fewer frogs mean more insects, and more insects mean more toxic pesticides. The wild harvesting extends to frogs and toads illegally caught for the million-dollar black market pet trade, risking not just the lives of the amphibians caught and transported but also raising the risk of imported diseases, viruses and parasites.

Frogs throughout the world, including our own native UK species, are most at risk however, from us humans and our need to take over and control the land and water. It is estimated that over 200 species of amphibian across the world

Natterjack toad.

have become extinct since the 1970s. The results of a survey published in 2004 determined that amphibians were more threatened than birds or mammals. Housing and agricultural development means vast tracts of land and wetlands are being swept up and concreted over, rivers diverted and lakes drained in order to make way for humans, housing and farms. Pollution and sewage in the waters, combined with use of pesticides and other chemicals, are further destroying the watery homes that remain for our amphibian friends. Much like otters and beavers, frogs are a very specific indicator of 'good' water, being very sensitive to water quality and displaced by pollution.

For our native species, the natterjack toad has had to contend with the disappearance of its coastal dunes for the tourism industry, and every year our common toads face the perils of road traffic on their migration to old breeding ponds. In addition, their preference for larger ponds and the reduction of these larger bodies of water thanks to development has resulted in an estimation that the number of common toads in the UK has declined by two-thirds in the last 30 years. Natterjack toads and pool frogs are now more reliant than ever on conservation efforts, designated sites and legal protections for their survival. Our common frog has adapted to make use of any watery garden feature it can find!

Thankfully, the efforts of conservationists mean that they are now being consulted in some building developments and wildlife corridors are being created. In cities and built-up areas these are tunnels or bridges built to link other wildlife habitats close by. In more rural settings these take the shape of hedgerows and woodlands, giving our amphibians and other wildlife, such as hedgehogs and badgers, access to alternative wildlife sites and wetlands as well as providing a degree of protection from predators and humans.

Pool frog.

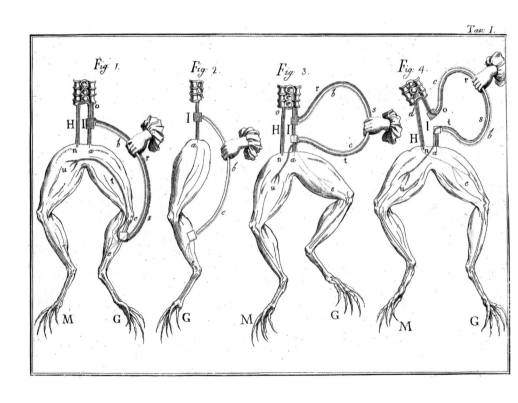

Luigi Galvani's experiments on the
sciatic nerve of frogs, 1792.

Under the Knife

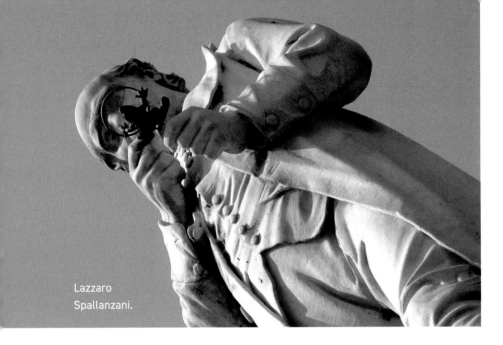

Lazzaro
Spallanzani.

Under the Knife

Our poor anurans have been a
fascination for the scientific and
medical fields for the last couple of
centuries. In 1780, Luigi Galvani, an
Italian physician based in Bologna,
conducted experiments using electrical
sparks on dissected frogs' legs,
causing them to twitch. His work
inspired Berzelius Volta to develop
the early battery, as well as Mary
Shelley's 1818 novel *Frankenstein*.
Just four years later and a couple of
hundred kilometres away in Pavia,
Lazzaro Spallanzani, a priest as well

as a biologist, was able to prove that fertilisation required both an ovum and sperm, using a male frog which he fitted with waxy 'trousers' to catch the sperm during amplexus. Spallanzani also used frogs in the first in vitro experiments.

Being easy to raise in large numbers and having many organs similar to other vertebrates, frogs quickly became the go-to for universities, research labs and hospitals.

As mentioned, the African clawed frog was bred in its thousands for pregnancy tests between 1940 and the early 1960s. Injected with the urine of a pregnant woman, the frog would very quickly begin to release eggs. The frogs were cheaper and gave a much speedier test result than using the previous method of rabbits and mice, as well as the 'convenience' of being able to use the frog repeatedly. However, once the pregnancy stick

African clawed frog.

test became more popular, thousands of frogs were simply released into the wild to fend for themselves. In true frog fashion, the African clawed frog has adapted to many different environments, although it is thought that they are, unfortunately, responsible for spreading the deadly fungal disease chytridiomycosis. Those African clawed frogs that weren't released into the wild went on to prove useful subjects, being one of the first species to be cloned (in 1973) and the only amphibians to be taken into space (in 1992). They continue to be used in stem cell research.

Whilst in the Middle Ages poisonous toads were regarded as evil, personifying the Devil and common familiars of witches, the use of their toxin was found to have pain-relieving effects and nowadays poisons created by toads and poison dart frogs are being studied carefully in the hope of developing more effective painkillers.

The hallucinatory effects of the yellow cururu toad (*Rhinella icterica*) are being studied in a quest for treatments for conditions like Alzheimer's and depression, and 5-MeO-DMT, a compound found in the venom of the Colorado River toad (*Bufo alvarius*) is currently being used in trials for the treatment of anxiety and even Post Traumatic Stress Disorder.

However fascinating and useful our wonderful anurans are, we still risk losing them to extinction. The Australian gastric brooding frogs (*Rheobatrachus silus* and *Rheobatrachus vitellinus*) are one such example. The female frog would brood its young in its stomach until the froglets were old enough, at which point they'd be 'vomited' out. Scientists were fascinated with the frog's ability to 'turn off' its secretion of gastric juice acid in the stomach whilst her young were growing.

Yellow cururu toad.

This curiosity wasn't enough to save the two species and, despite only being 'discovered' in the 1970s, both were classified as extinct by the 1980s, thought to have been wiped out by chytrid.

With all the other threats that frogs are currently facing, from habitat destruction, poor water quality and a wide-range of predators and disease, we need to do our level best to ensure their protection and future.

Frog Spotting

Frog Spotting

My favourite time to spot the frogs in my pond is when the sun has been warming up the exposed bits and I'm treated to frogs of all sizes basking and soaking up the heat in and around the water.

Some cling to the edges of the pond and others sit half-submerged with just their heads poking out, keeping an eye on things. My other favourite time is in the evenings, sneaking out with a torch and counting up the numbers. I'm very lucky that there was a pond when I arrived at this property, but the pond was very different back then, no plant life and a lot of fish! Over the years, I've added heaps of plants in and around the pond and encouraged ferns and lots of plant pot hidey spots for frogs. Sadly, a heron gobbled up most of my fish, but that has resulted in a much friendlier environment for the emerging tadpoles!

It did seem a bit 'build it and they will come' when, as a birthday present, my partner Sean bought a pre-formed 'patio' wildlife pond for me. I keep a pond pump and fountain running in the main pond, but we wondered if frogs might prefer the still water in a simple wildlife pond. Sure enough, it was the place the frogs first laid their spawn. Cutting that pond back in the autumn is always filled with lots of joyous whoops and hollers as frogs, big and small, leap out and dive into the bigger pond right next to it, waiting for me to be done with the autumn clear out and just clear off! I also ensure I leave all cuttings and plants near the pond for a few days to dry out before green-

bagging them, that way any creatures or froglets caught up in the clear out have a chance to climb out while it is quieter and make their way back to the pond.

If you've the space and means to dig a pond in your garden – however big or small – it will be a welcome oasis for all sorts of wildlife, not least of which will be our amphibian friends. Even small containers like old sinks and barrels make good homes; just make sure there's plenty of options for the froglets and frogs to be able to get in and out, as well as means for hedgehogs to escape too, should they fall in. My partner's 'pond' for several years was his children's discarded plastic sandpit which he left partially covered with its lid. He revelled in peeking under the lid from time to time and greeting its inhabitants.

Wood piles, rockeries and spaces under sheds are perfect spots for our toady friends, but be sure not to bother them too much and allow them to amble away when they've decided it's time to go. Try not to pick up a frog unless absolutely necessary – that is, to move it away from imminent danger – as our dry skin can affect the delicate water balance in theirs, causing the water to diffuse out. In summer, when the warm summer rains have encouraged my froglets to leave 'The Mother Pond', I use a pint glass to rescue those errant adventurers that have somehow hopped into the house. A handy tip: frogs can't jump backwards, so placing the pint glass in front of them and moving something behind them will encourage a forward leap.

As tempting as it is, don't share tadpoles between ponds, particularly across counties. The risk of spreading diseases amongst populations just isn't worth it and, as we've seen, the diseases are quite awful and very easily spread.

There are many wildlife groups who arrange visits to known breeding spots in early spring or evening trips during autumn where you've a chance to watch frogs and toads. They will also likely offer opportunities to join invaluable Toad Patrols during which you can help protect and record the numbers of toads on their annual migration. Remember, however, that to a frog and toad you're a BIG predator, so step carefully and as slowly and quietly as possible. In my garden, even stepping as gently as I can, I see several heads disappear underwater as I approach. Standing still and quiet for a few minutes usually rewards me with those heads poking back up from under the water to have a good old peer at me, as if saying, 'You again?!'

Teaming up with other frog spotters!

Herpetology is the study of amphibians, like frogs and toads, as well as reptiles like snakes and lizards. The word is derived from the Greek *herpetón*, meaning 'creeping animal'. Across the whole of the UK, you'll find a variety of groups and charities dedicated to studying and conserving our frogs.

Amphibian and Reptile Groups of the UK (ARG UK) is the umbrella organisation connecting and supporting a wide network of independent Amphibian and Reptile Groups (ARGs) across the country. Alongside preservation and conservation work, they also strive to educate and raise awareness of the wide range and diversity of our native species. Importantly, they encourage collaboration between the ARG groups

which facilitates the collation and sharing of information on local species. On their website you can not only record your own sightings but also look at all the data currently recorded and even view it on an interactive map so you can see what's happening in your area. They are always on the lookout for volunteers for their various projects. For more information visit www.arguk.org.

Beds Reptile and Amphibian Group is the affiliated ARG group in Bedfordshire and they organise occasional outings, training sessions and workshops for those with an interest in local reptiles and amphibians. They too are on the lookout for volunteers and recorded sightings to accurately map the numbers and diversity across the county. The RSPB's Lodge nature reserve in Sandy is the only inland habitat for natterjack toads thanks to a successful reintroduction programme in the 1980s.

Marsh toads and midwife toads are also happy to call Bedfordshire home. Find out how you can get involved with a local group at www.groups.arguk.org.

Froglife is a registered charity operating across the UK dedicated to the conservation of amphibians and reptiles. They focus on improving habitats as well as education outreach, research and campaigning. They will also help you connect with your local Toad Patrol groups. Find out more at www.froglife.org.

The second plague of Egypt: frogs. Picture from the popular Bible encyclopedia of
Archimandrite Nikiphor, 1891.

Frogs in Myth and Legend

The 'magical' metamorphosis of the frog, its ability to shed its skin as well as living in both water and on land, has propelled the frog into many myths and legends across several cultures. In some cultures, the use of toad toxins is thought to play a part in some shamanistic and religious transcendental rituals.

During ancient life around the Nile, frogs were revered as symbols of life and fertility, especially as their appearance heralded the rains, the waters rising and the flourishing of crops. There was even a frog goddess, Hequet (or Heket), 'she who hastens the birth'. Midwives became known as 'servants of Hequet' and good luck charms and amulets in the shape of frogs were carried and worn by women during childbirth.

Pliny the Elder

Pliny the Elder (Gaius Plinius Secundus, 23-79 AD) was not only a naval and army commander but also a keen naturalist and published 10 volumes of the *Natural History* before his death. Some of his suggested remedies, however, are decidedly odd! The best cure for quartans (a malaria-like fever): a frog, worn as an amulet with its claws taken off, or a bramble toad, if its liver or heart is worn as an amulet in a piece of ash-coloured cloth.

He believed that frogs were reborn from the muddy rivers during the annual spring rains and also mentions storms of frogs and fish. Given that frogs do indeed reproduce around the time of spring rains and the meteorological phenomenon of 'raining frogs' has been documented, Pliny the Elder clearly had a keen eye for the craziness of nature!

Faience Frog from the House of the Silver Wedding, Pompeii, 1st Cent. AD,
Naples Archaeological Museum.

Ranarum, Ciniphum, Muſcarum funditur agmen. *Et terra natis corporibuſque nocent.* Exod. 8.

The Plague of Frogs by Jan Sadeler I,
after Maarten van Cleef, 1585.

Amphibians seemed to have a particularly onerous time in the Christian teachings, beginning with the plague of frogs, as recounted in Exodus as the second of the ten plagues wrought upon the Egyptians to force the freedom of the Israelites. This plague promised the Nile swarming with frogs, '...which will come up and go into your house and into your bedroom and on your bed... and into your ovens and into your kneading bowls'. The sudden appearance of hundreds of frogs in wetlands simply naturally making their way out could, very likely, be seen as a plague!

Very quickly, frogs and toads came to be seen as satanic and evil. Given that it is likely that early man was already acquainted with the toxicity of toads – either in a hallucinogenic form or even fatal – frogs and toads became synonymous with bad luck and even witchcraft during medieval times. A frog walking across your foot was bad luck, and one turning up inside your house meant an imminent death. Folk attempted to deter the creatures with 'magical' stones and would spit on them if spotted.

Toad stones

As far back as the 12th century there were references to 'toad stones', a magical jewel toads carried around in their heads. Even Shakespeare made reference to this in his 17th-century play *As You Like It*, where the exiled Duke Senior says: 'Sweet are the uses of adversity / Which, like the toad, ugly and venomous/Wears yet a precious jewel in his head'. Alas, these 'precious' jewels turned out to be the fossilised teeth of extinct fish dating back to the Jurassic Period, but they were round and button-shaped and could be worn in small bags hung around the neck or even placed in jewellery.

This stone eventually became classified as bufonite – after the Latin word for toad, *bufo*. Macabrely, Edward Topsell, who collected details of various creatures and beasts from around the world, suggested in his book *History of Four-Footed Beasts and Serpents*, published in 1658, that a toad stone was best collected from a live old toad's head by sitting it on a piece of red cloth.

Shakespeare

Shakespeare also mentioned frogs in *Macbeth*, written a few years after *As You Like It*. On this occasion, however, Shakespeare linked them with witches, specifically the Second Witch, who refers to her 'familiar' – or assistant – as Paddock. Paddock and also puddock are, respectively, old English and Scottish words for frog.

Urshula Kemp

In 1582, Urshula Kemp from St Osyth in Essex was tried and hanged for witchcraft. In her confession she admitted to having a familiar in the form of a black toad named Pygine.

'Sam Appleby, Horseman' by John Reibetanz

In the folklore of the Fens of East Anglia and Lincolnshire, talk of the 'Toadman' who could 'both heal and hex' horses appears to have lasted until as recently as the 1930s. The Toadman had the ability to calm a raging horse, or turn a tame one wild so that no one else could calm it. In the days of ploughing and when all movements between towns and villages were horse drawn, this was some ability. In his 1979 poem 'Sam Appleby, Horseman', American/Canadian poet John Reibetanz describes the ritual for gaining this knowledge.

I hung the limp toad up to dry
Overnight on a blackthorn tree, like Shecky said,
Then stuffed it in an anthill for a month,
And by the full moon's light pulled out a chain
Of bones, picked clean.
Then came the tricky part.
You carry the skeleton to a running stream
To ride the moonlit water, but you dare not
Take your eyes off it till a certain bone
Rises and floats uphill against the current;
Then grab this bone— a little crotch bone it is,
Shaped like a horse's hoof—and take it home,
Bake it and break it up into a powder:
The power's in the powder.

St Ulphia

There was one Christian, however, who seemed fond of frogs, their patron saint, St Ulphia. Legend has it that while already living as a hermit as a young girl, Ulphia encountered a group of men driving frogs from a river intent on killing them, as their croaking was proving a nuisance to a local priest. Ulphia promised to protect the frogs and they hid under her robes and clung to her legs as she walked back to her home on the River Noye in the Hauts-de-France region of northern France. There, the frogs were free to live in peace, but she bade them to be quiet lest they be discovered or, indeed, interrupt her own prayers and sleep. According to a biographer of saints in the 19th century, the frogs around the River Noye are indeed quieter but, even more surprisingly, turn out to be louder if moved to another area.

Left: Statue of Ulphia at Amiens Cathedral.

Chinese culture

In Chinese culture and *feng shui* traditions, the 'money frog' is seen as representing wealth and prosperity and should be encouraged. Should he appear during a full moon, good fortune is foretold. The figurines are commonly depicted as a large frog sitting on a pile of coins with a coin in its mouth. *Feng shui* principles suggest he should sit facing the main door of the house, encouraging wealth to enter. The original story of the 'money frog' goes back to a monk named Liu Haichan, who is thought to have studied all the spiritual practices needed to achieve an incredibly long life – even immortality – and is usually depicted carrying his three-legged toad or frog.

Japanese folklore

The corresponding tale in Japanese folklore tells of Kō Sensei, also known as Gama Sennin, 'Immortal with his Toad', who, with warty features much like his travelling companion, wanders the land imparting wisdom.

Right: Gama, the Immortal with Toad, a Japanese wooden figure from the first half of the 19th century.

Mexican culture

The Olmecs, an ancient civilisation which inhabited the area south of modern-day Mexico City between 1500 and 400 BC, worshipped toads as a sign of fertility and were fascinated by their ability to eat their own discarded skin. This they took as a symbol of rebirth after death and incorporated the theme into their shamanic rituals.

The Maya Earth Lord, Tlaltecuhtli, was often represented as a giant toad-like creature. They also regarded frogs as signs of fertility and that their croaking heralded the start of the rainy season – chirping and whistling along with the croaking was part of the rain rituals. Today, the güiro percussion instrument used around South America is sometimes made in the form of a frog or toad, with the noise created when the stick is stroked across the notched back. When not in use, the stick is stored safely in the frog's mouth.

In other parts of the bygone era of the Mexicas there was the more practical practice of catching frogs and toads and eating them as a good source of protein (tadpoles could even be bought at markets for raising your own). The toad poison, bufotenine, was used in hallucinogenic practices by priests.

Francisco Hernández de Toledo (1514-1587), a naturalist and surgeon who travelled from Spain to what is today Mexico on the orders of the Spanish king, spent seven years studying and collecting specimens from the area. Apart from confirming that the amphibians were quite tasty, he also noted that dried frog intestines were used in treating kidney stones and toads were useful as diuretics and even blood purifiers.

Indian culture

Built around 1870, the Frog Temple, or Manduk Mandir, in Uttar Pradesh, northern India probably gives a fairly clear indication of how frogs are considered in Indian culture. Legend states that the king of the area, Bakhat Singh, received a blessing from a frog and he and his progeny lead prosperous lives from then on. They still maintain the management of the shrine, and the well in front of the frog's mouth is said to never dry up.

Even today, 'frog weddings' are arranged as part of rain rituals in the belief that a marriage between two frogs would please the rain god, Indra. In the Mandukya Upanishad (or Upanishad of the Frog), it is told that through meditation and self-reflection one could leap like a frog to salvation.

Mandauk Mandir, or the Frog Temple,
Uttar Pradesh, India.

Picture Book of Crawling Creatures, Metropolitan
Museum of Art, New York, c. 1823.

Frogs in Art
and Literature

With their big bulgy eyes and long legs, frogs often feature in a cartoony fashion. Kermit the Frog from *The Muppets*, the backing band in Paul McCartney's 1984 short animated film "We all stand together", the Frogger arcade game and even the CGI musician, Crazy Frog – frog cartoons abound. Needless to say, there are plenty of frogs in literature too.

Upon The Frog
John Bunyan (c.1628–1688)

Born in Elstow, Bedfordshire, John Bunyan is best known for *The Pilgrim's Progress from This World, to That Which is to Come,* published in 1678. In this poem, Bunyan is quite unflattering about the frog, likening her to a hypocrite.

The frog by nature is both damp and cold,
Her mouth is large, her belly much will hold;
She sits somewhat ascending, loves to be
Croaking in gardens, though unpleasantly.
Comparison.
The hypocrite is like unto this frog,
As like as is the puppy to the dog.
He is of nature cold, his mouth is wide
To prate, and at true goodness to deride.
He mounts his head as if he was above
The world, when yet 'tis that which has his love.
And though he seeks in churches for to croak,
He neither loveth Jesus nor his yoke.

Right: Illustration from *Historia naturalis ranarum nostratium* by August Johann Rösel von Rosenhof, 1758.

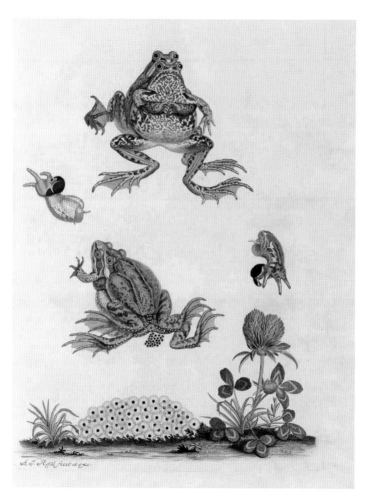

Frog House is an example of Art Nouveau
architecture in the city of Bielsko-Biała, in
southern Poland's Silesia Province.

The Frog Prince and the Frog Princess

The Frog Prince, or Iron Henry, was part of the first collection of tales published in 1812 by the Brothers Grimm, Jacob and Wilhelm, academics who collected and published folk stories. It is a tale of an enchanted frog 'transforming' into a handsome prince.

The story tells of a young princess playing with a favourite golden ball near a spring of water. She threw the ball too high, missed the catch and the ball splashed into the water, sinking immediately. She began to cry. A frog popped his head out of the water and asked the princess why she was so sad. The princess told him about her lost ball and said she'd give anything to get it back. The frog said he wouldn't want any jewels or pearls as a reward, but if she would love him and let him live with her and eat from her golden plate and sleep on her pillow, he would

find the ball and return it to her. The princess thought the frog was being silly, that he wouldn't even be able to follow her home, so agreed to his conditions. With that the frog slipped under the water, came back up with the ball in his mouth and tossed it onto the edge of the spring. The spoiled princess was delighted and thought nothing of snatching up the ball and running away as fast as she could, ignoring the frog's calls after her.

That evening, as the princess sat down to dinner with the king and queen, there was a splashing outside the door, followed by a knock and a voice calling for the princess. The princess opened the door and, to her horror, there sat the frog, asking to come inside. With a shriek, the princess slammed the door shut and ran back to the table. The king asked her what's frightened her so and the princess told her father the whole story. 'Well,' said the king, 'you have given your word so you must keep it. Go and let him in.' And so the Princess let in the frog and he hopped over to the table. He asked the princess to lift him onto the table and let him sit next to her, which she did grumpily. He asked her to push her golden plate toward him so that he could eat from it, she did so and the frog ate as much as he could. Afterwards, he announced he was tired and would she take him to sleep on her pillow. Again, the princess obliged.

The following morning the frog woke at dawn, hopped off her pillow, down the stairs and out of the palace. The princess was relieved and thought that was the end of it, but not so. The following evening, the frog tapped at the door and asked to be let in. Grudgingly, she did so and allowed him to eat from her plate and sleep on her pillow again and the frog hopped away as dawn broke. The third night the same thing happened. But the following morning, instead of a frog on her pillow, there

stood a handsome prince gazing at her lovingly. He told her that he'd been enchanted by a spiteful fairy who had changed him into a frog. The spell could only be broken by a princess who would let him into her home, eat from her plate and sleep on her pillow for three nights. Now that she had broken the spell, he would like to marry her and take her home to his kingdom. And they lived happily ever after.

The reference in the story's title to 'Iron Henry' is the prince's loyal man-servant who, fearing his heart would burst with sadness when his beloved prince was put under the spell, had three iron bands fixed around his chest. The sight of the spell lifted from his master made his heart swell so much with happiness that the iron bands burst.

There have been many variations on the original tale since first published, Including a 1971 television special,

Sumo wrestling toads by Hōson, c. 1930.

The Frog Prince, with the world's most famous frog Muppet, Kermit, as the narrator. Our Frog Prince turned out to be King Harold, Princess Fiona's father, in the 2004 animated film *Shrek 2*, and Disney produced its animated variation in *The Princess and The Frog,* released in 2009.

A variation on the original old folk tale has an equally old version in which the roles are reversed and the female character of the story is the enchanted frog: The Frog Princess or The Tsarevna Frog.

Folk Tales From The Russian

One such telling appears in *Folk Tales From The Russian*, retold by Verra Xenophontovna Kalamatiano De Blumenthal, first published in 1903.

There once was a Tsar who decided it was time his three sons should marry. He instructed all three to fire an arrow into the air and they would find their bride where the arrow landed. The first son's arrow landed in the courtyard of a nobleman's home and he immediately proposed to the nobleman's daughter. The second son's arrow landed on the porch of a rich merchant's home and the prince offered his hand in marriage. The third son, Ivan, sent his arrow off and it landed in the swamp right next to a small frog. Ivan carried the frog home to his father, who insisted Ivan marry the frog as, clearly, that was his destiny.

Before the weddings, however, the father tests the brides-to-be with tasks. The first night's task is to spin cloth into a shirt fit for a king. Ivan is convinced his little frog will not be up to the task but, instead, she produces the most magnificent shirt the following morning, far superior to the other two. The next test is to bake a loaf of bread. This time, however, Ivan is curious to see how his bride-to-be had managed her last task and hides. To his surprise, he watches a beautiful woman undress out of her frogskin, mix the dough and fling it out of the window onto a magical wind, before dressing back into her frogskin. The following morning, the frog's bread is deemed the best in the land and the other two inedible.

Ivan is transfixed by the beautiful woman and resolves to destroy the frogskin at the first available opportunity.

The next task for the prospective brides is to dance at the upcoming banquet to prove they're fit for court. The beautiful maiden herself turned up at the banquet and dazzled her audience with an enchanting display, whilst the other two maidens were mocked and laughed at for trying to copy the frog-maiden's magic and failing. Ivan was nowhere to be seen.

The frog-maiden returned to her chamber just in time to watch Ivan throw her frogskin onto the fire.

She was distraught and wept, asking why Ivan could not trust her and, with that, she flew out of the window on the magical wind.

Ivan was heartbroken and sought out the wisest counsellor, who listened to the story and said that the frog-maiden was none other than Vasilissa, a powerful enchantress, whose father feared she was becoming more powerful than him and, in his rage, put her under a spell that could only be broken by a prince who would marry a frog. The counsellor instructed Ivan to travel to Baba Yaga's house, the old witch who lived in a house on chicken legs deep in the northern forests. And so he did.

Baba Yaga was not impressed with Ivan at first, but seeing how desolate he was she took pity on him. She instructed him to hide and when Vasilissa returned to visit, Ivan was to grab a hold of her and not let go.

Well, it was quite a chase, as Vasilissa quickly turned back into a frog and was able to jump about and slip out of his grasp quite easily. Eventually, Ivan gave up and slumped onto the floor lamenting his mistake in burning her frogskin. Vasilissa saw that he was truly remorseful and truly loved her and allowed Ivan to pick her up. He tucked her into his shirt, right next to his heart, and Baba Yaga gifted him a magic carpet to fly them both home.

The day of the three weddings arrived and Ivan proudly held his frog bride in his hand and kissed her on her wide mouth. With that, the spell was broken and Vasilissa, the wise and beautiful, was restored to her true form.

A Frog's Fate by Christina Rossetti (1830–1894)

Poet Christina Rossetti was born in London and was the younger sister of the painter Dante Gabriel Rossetti, often appearing as his model. In this poem, published in 1885, she tells the woeful tale of a frog wanting to explore the big world outside of his small pond.

Contemptuous of his home beyond
The village and the village-pond,
A large-souled Frog who spurned each byeway
Hopped along the imperial highway.
Nor grunting pig nor barking dog
Could disconcert so great a Frog.
The morning dew was lingering yet,
His sides to cool, his tongue to wet:
The night-dew, when the night should come,
A travelled Frog would send him home.
Not so, alas! The wayside grass
Sees him no more: not so, alas!

A broad-wheeled waggon unawares
Ran him down, his joys, his cares.
From dying choke one feeble croak
The Frog's perpetual silence broke: –
'Ye buoyant Frogs, ye great and small,
Even I am mortal after all!
My road to fame turns out a wry way;
I perish on the hideous highway;
Oh for my old familiar byeway!'
The choking Frog sobbed and was gone;
The Waggoner strode whistling on.
Unconscious of the carnage done,
Whistling that Waggoner strode on –
Whistling (it may have happened so)
'A froggy would a-wooing go.'
A hypothetic frog trolled he,
Obtuse to a reality.
O rich and poor, O great and small,
Such oversights beset us all.
The mangled Frog abides incog,
The uninteresting actual frog:
The hypothetic frog alone
Is the one frog we dwell upon.

Christina Rossetti.

The Frog by Hilaire Belloc (1870–1953)

A very prolific writer, Hillaire Belloc also wrote poetry for children. *The Bad Child's Book of Beasts* was published in 1896, with illustrations by Basil Temple Blackwood. In the poem, despite suggesting you ought not call the frog names, the author does appear to be poking fun, listing a variety of silly names for frogs and even suggesting those that like frogs are rare and even lonely!

Be kind and tender to the Frog,
And do not call him names,
As 'Slimy skin,' or 'Polly-wog,'
Or likewise 'Ugly James,'
Or 'Gape-a-grin,' or 'Toad-gone-wrong,'   Or 'Billy Bandy-knees':
The Frog is justly sensitive
To epithets like these.
No animal will more repay
A treatment kind and fair;
At least so lonely people say
Who keep a frog (and, by the way,
They are extremely rare).

The Tale of Mr. Jeremy Fisher by Beatrix Potter

First published in 1906, the story follows the adventures of Mr. Fisher, the frog, who lived in a damp, 'slippy-sloppy' house and liked to get his feet wet. One rainy day, he decides to go fishing and plans to invite his friends – Mr. Alderman Ptolemy Tortoise and Sir Isaac Newton (a newt) – around for tea, should he catch enough fish.

There follows a most adventurous day in which Mr. Fisher accidentally catches a spiny stickleback which flapped and flopped around in the boat and prickled and snapped at poor Mr. Fisher before jumping back into the water. But, as Mr. Fisher is nursing his sore fingers, he's suddenly caught by a huge trout, who drags him into the water. Fortunately, the trout decides the taste of Mr. Fisher's mackintosh is really unpleasant and spits him out. Mr. Fisher dashes home minus his galoshes, rod and basket vowing never to go fishing again. His two friends do pop round for tea however, with Mr. Tortoise bringing a salad and Mr. Fisher cooking up roasted grasshopper with lady bird sauce for himself and Sir Newton.

Mr. Jeremy Fisher popped up again around 1910 in a series of miniature letters Ms. Potter wrote to child fans. One child suggested Mr. Fisher should

get married but in one of the four letters back to the child, Mrs Tiggy-winkle, a hedgehog washer woman that featured in an earlier Potter book, vigorously declined any notion of marrying Mr. Fisher on the basis of his 'slippy-sloppy' house and her not being at all keen on starching his cravats.

Toad of Toad Hall –
Wind in the Willows

In Kenneth Grahame's much-loved 1908 book, rambunctious Toad (or 'Toady' to his friends) is very wealthy, very generous and very easily distracted by his faddish interests and pursuits. His newest passion is motorcars and it's getting him into a lot of trouble, as he keeps crashing them! Rat, Mole and Badger travel to Toad Hall to try and convince him to give up this crazy and dangerous hobby, but Toad escapes by knotting sheets from his bed and sliding down them out of his window. His friends resolve to remain at Toad Hall until Toad returns and they'll try again. But Toad, however, has stopped for a spot of lunch at the pub and spots a motorcar pull up with a group of people popping into the pub as well. Unable to contain himself, Toad steals the car and tears away recklessly. It is not long before

Above: A used postage stamp printed in Britain showing characters from *The Wind in the Willows*, 1979.

he is caught. The magistrate hands him a prison sentence of 20 years and Toad finds himself locked up in a dungeon. Oh, poor Toad! Fortunately, the gaoler's daughter takes pity on him, despite all his histrionics, and they become friends. Not long afterward, she helps him escape the prison by dressing him as a washerwoman. After several adventures, Toad finally bumps into his old friend Ratty, only to hear that his beloved Toad Hall has been taken over by the stoats and weasels, known as the Wild Wooders. In one last adventure, Rat, Mole, Badger and Toad carry out a cunning plan of sneaking into Toad Hall, surprising the interlopers and chasing them all off. Through all his adventures, Toad has realised how silly he's been and appreciates his wonderful, loyal friends. He compensates all those that had helped him along the way and settles down to a quieter life.

The Fairy Frog by Gertrude Landa (1892–1941)

Writing under the pseudonym Aunt Naomi, Landa tells the tale of a magical frog in *Jewish Fairy Tales and Legends*, published 1919.

Hanina, the only grown son of two elderly parents, arrives one day to find them both dying. As a final wish, the father instructs Hanina to mourn for the customary seven days, but on the eighth day to travel to the market and buy the first thing that is offered to him, no matter what it is or what it costs. This, promises the father just before passing, will bring Hanina good fortune.

Hanina sadly buried his parents and mourned. He was a good son and, as instructed, on the eighth day Hanina went to the marketplace, curious about what would happen. The market was busy with bustling traders and Hanina

Illustration by an unknown artist, published in *Jewish Fairy Tales and Legends* by Aunt Naomi, Bloch Publishing Company, 1919.

did not know where to start. But very soon, an elderly man carrying a silver casket with strangely designed engravings approached Hanina. The old man offered the casket to Hanina and urged him to buy it. 'It will bring thee great fortune,' said the man. Hanina asked what was in the casket and the old man said he would not tell him no matter how many times he asked. 'How much does it cost?' enquired Hanina. 'A thousand gold pieces,' smiled the old man.

A thousand gold pieces was a tremendous amount of money, but Hanina respected his father's dying wish, paid the old man the money and went home to his wife with the strange casket.

Hanina placed the casket on his table and, while his wife stood next to him, they carefully lifted the lid. Within the casket was a smaller casket. Gently, they lifted that lid too. Out hopped a

frog. Both Hanina and his wife were sorely disappointed but, being good people, they fed the frog. The frog gobbled up all the food they gave it, so they gave it some more; and the frog gobbled that up too. After eight days, the frog was already so big that Hanina built a special cabinet for the frog. But soon, after eating more and more food, Hanina had to build a shed for it. Still the frog continued to eat and eat; before long Hanina and his wife were having to sell off their possessions to buy food for the frog, oftentimes going without themselves.

One day, Hanina's wife could bear it no more and burst into unhappy tears. Suddenly, the frog, who was now larger than a full grown man, spoke to Hanina's wife. 'You and Hanina have treated me kindly. Ask me and I shall grant your wish.' Hanina's wife sobbed, 'Please, give us food.'

'Your food is here,' said the giant frog, and with that there came a knock at the door and a huge basket of food was there for them. The frog asked Hanina for his wish. Hanina decided that a talking frog must be very wise indeed and asked for all the knowledge of man. With that, the frog wrote out all the knowledge of man on strips of paper. The frog then handed the strips of paper to Hanina and bade him to swallow them all. Hanina did as he was told and, as he did, he suddenly acquired all the knowledge of the world – even the language of the animals. Hanina was regarded as the wisest of the wise.

Not long afterwards, the frog spoke again to Hanina and his wife. 'You have treated me with great kindness and it is now time for me to repay that kindness. It shall be a great reward. Come, follow me.' And with that, Hanina and his wife followed the enormous frog into the woods.

When they were deep inside the great forest, the frog croaked out loudly and summoned all of the creatures of the woods and rivers. 'Come, inhabitants of the trees and caves and waters. Bring precious stones from the depths and bring roots and herbs!' And with that, all the creatures, from the tiniest insects to the birds flying around, began to pile jewels and herbs at the feet of bewildered Hanina and his wife. The frog spoke again, 'All these riches are yours – precious stones will make you rich and the roots and herbs will cure all diseases. You obeyed the wishes of your dying father and you treated me well, never questioning me. Now this is your reward.'

Hanina and his wife were amazed and thanked the frog profusely. 'But, may we ask who you are?'

'I am the fairy son of Adam and I can assume any form. Farewell.' And with that, the enormous frog got smaller

Frog by Kate Wyatt.

Frog prince, Irish stamp, 1998.

and smaller and smaller until he was no bigger than an ordinary frog, jumped into the river with a splash and disappeared.

Hanina and his wife went home with their gifts. They lived happily for many, many years and were known far and wide for their wealth, their wisdom, their healing and their charity.

The Wishing Fairy's animal friends

The Peacock and the Wishing-fairy and Other Stories is a delightful collection of short stories by Corinne Ingraham, published in 1921, with illustrations by Dugald Stewart Walker. In much the same vein as Rudyard Kipling's *Just So Stories*, several Wishing Fairy books were published which tell the tales of how the peacock got its glorious tale and the zebra got its stripes. The Wishing Fairy's name is Stella and she lives in a lily house at The End of the Earth. She's assisted by Brownies and animals can visit the Get-little-pond or the Get-big-pool, to match their size to the tiny fairy when they visit her, in the hopes that she will make their dreams come true.

We first meet Mr and Mrs Frog when they accompany Turtle to visit Stella at The End of the Earth. Turtle hoped the Wishing Fairy could think up a way to help him hide from predators like Pelicans. After a bit of time under the thinking cap, Stella created Turtle's hard shell and he was able to simply pull his head and feet in whenever he was in danger. Everyone was delighted!

Mr and Mrs Frog were soon to visit Stella, the Wishing Fairy, for themselves. After Mr Frog got into a big fight with another frog over a missed fly, Mr Frog lamented that he wished his mouth was bigger to help him catch flies. He announced to Mrs Frog that he was off to The End of the Earth to ask the Wishing Fairy for his wish. Mrs Frog thought that was a splendid idea and hopped along as well. Very soon, the Wishing Fairy had granted their wish. Mr and Mrs Frog hopped happily all the way home and, before long, all their frog friends were very curious about their new large mouths! An enormous moth flew by and, quick as a flash, Mr Frog snapped it up. The other frogs were amazed and suddenly every frog in the pond had a wish – they all had the same wish – and they all hopped off to The End of the Earth to ask Stella, the Wishing Fairy, to grant them their wish of a big mouth. Stella made all their wishes come true and they were all very happy, and now you know how the Frog got his big mouth and why.

The Puddock
John M. Caie (1878-1949)

In *The Kindly North: verse in Scots and English*, published 1934, John Caie enjoyed writing poetry about life and nature in the north-east of Scotland. This poem is written in his native Doric dialect.

Meaning of unusual words:

puddock – frog
hurdies – buttocks
seggs – yellow iris
gapit – gaped open
gin – if
thrapple – throat
wame – stomach
chiel – child
blaw – boast
nabbit – grabbed
syne – afterwards
peer – poor

A puddock sat by the lochan's brim,
An' he thocht there was never a puddock like him.
He sat on his hurdies, he waggled his legs,
An' cockit his heid as he glowered throu' the seggs.
The bigsy wee cratur' was feelin' that prood,
He gapit his mou' an' he croakit oot lood:
"Gin ye'd a' like tae see a richt puddock," quo' he,
"Ye'll never, I'll sweer, get a better nor me.
I've fem'lies an' wives an' a weel-plenished hame,
Wi' drink for my thrapple an' meat for my wame.
The lasses aye thocht me a fine strappin' chiel,
An' I ken I'm a rale bonny singer as weel.
I'm nae gaun tae blaw, but th' truth I maun tell-
I believe I'm the verra MacPuddock himsel'." ...

A heron was hungry an' needin' tae sup,
Sae he nabbit th' puddock and gollup't him up;
Syne runkled his feathers: "A peer thing," quo' he,
"But – puddocks is nae fat they eesed tae be."

A Frog He Would A-wooing Go

No one is entirely sure who originally wrote the ditty, but its first published appearance was in Scotland in 1549 under the name 'The Frog came to the Myl dur'. It has since appeared in various forms including 'Frog Went A-Courting'. The version published in 1883 by Frederick Warne & Company features illustrations by Chester-born artist Randolph Caldecott (1846-1886).

The story tells of Frog, who is keen to woo Miss Mousey and pays her a visit, along with his friend, Rat. It all ends a bit unfortunately, with Cat and her kittens crashing into Miss Mousey's house and attacking Miss Mousey and Rat. Frog manages to escape the ambush, only to stumble upon Duck, who gobbles him up!

A Frog he would a-wooing go,
Heigho, says Rowley!
Whether his Mother would let him or no.
With a rowley-powley, gammon and spinach,
Heigho, says Anthony Rowley!

So off he set with his opera-hat,
Heigho, says Rowley!
And on his way he met with a Rat.
With a rowley-powley, gammon and spinach,
Heigho, says Anthony Rowley!

'Pray, MR. RAT, will you go with me,'
Heigho, says Rowley!
'Pretty MISS MOUSEY for to see?'
With a rowley-powley, gammon and spinach,
Heigho, says Anthony Rowley!

A Frog He Would A-Wooing Go by
Randolph Caldecott, 1883.

Now they soon arrived at Mousey's Hall,
Heigho, says Rowley!
And gave a loud knock, and gave a loud call.
With a rowley-powley, gammon and spinach,
Heigho, says Anthony Rowley!

'Pray, Miss Mousey, are you within?'
Heigho, says Rowley!
'Oh, yes, kind Sirs, I'm sitting to spin.'
With a rowley-powley, gammon and spinach,
 Heigho, says Anthony Rowley!

'Pray, Miss Mouse, will you give us some beer?'
Heigho, says Rowley!
'For Froggy and I are fond of good cheer.'
 With a rowley-powley, gammon and spinach,
Heigho says, Anthony Rowley!

'Pray, Mr. Frog, will you give us a song?
Heigho, says Rowley!
'But let it be something that's not very long.'
With a rowley-powley, gammon and spinach,
Heigho, says Anthony Rowley!

'Indeed, Miss Mouse,' replied Mr Frog,
Heigho, says Rowley!
'A cold has made me as hoarse as a Hog.'
With a rowley-powley, gammon and spinach,
Heigho, says Anthony Rowley!

'Since you have caught cold,' Miss Mousey said.
Heigho, says Rowley!
'I'll sing you a song that I have just made.'
With a rowley-powley, gammon and spinach,
Heigho, says Anthony Rowley!

But while they were all thus a merry-making,
Heigho, says Rowley!
A Cat and her Kittens came tumbling in.
With a rowley-powley, gammon and spinach,
Heigho, says Anthony Rowley!

The Cat she seized the Rat by the crown;
Heigho, says Rowley!
The Kittens they pulled the little Mouse down.
With a rowley-powley, gammon and spinach,
Heigho, says Anthony Rowley!

This put Mr. Frog in a terrible fright;
Heigho, says Rowley!
He took up his hat, and he wished them good night.
With a rowley-powley, gammon and spinach,
Heigho, says Anthony Rowley!

But as Froggy was crossing a silvery brook,
Heigho, says Rowley!
A lily-white Duck came and gobbled him up.
With a rowley-powley, gammon and spinach,
Heigho, says Anthony Rowley!

So there was an end of one, two, and three,
Heigho, says Rowley!
The Rat, the Mouse, and the little Frog-gee!
With a rowley-powley, gammon and spinach,
Heigho, says Anthony Rowley!

Photo Credits and Artworks

Front cover: Tanya Ware.

Back cover left to right: Tanya Ware, Sean Rudkin, Steve Owen, Tanya Ware.

Introduction
Tanya Ware: page 4.
Jo Byrne: page 7.
Andy Coventry: page 8.

Frog Physiology
Jo Byrne: pages 10, 15, 50.
Sean Rudkin: pages 13, 20-21, 22, 43, 45 (top right, bottom left and bottom right), 47-49.
Katya/Flickr: page 14.
Shutterstock: page 19.
Steve Owen: page 18.
Tanya Ware: pages 25, 26, 30-31, 35, 37, 38-39, 42, 45 (top left).
Charles J Sharp/Wikimedia: page 29.
John Spiers: pages 32, 34, 54.
Andrew Bone/Flickr: page 33.
hdwallpapersfreedownload.com: page 40.
Andy Coventry: pages 46, 51, 53, 55.

UK Native Frogs and Toads
Steve Owen: page 56.

Sean Rudkin: page 58.
Tanya Ware: pages 59, 60-61.
Viridiflavus/Wikimedia: page 63.
Sean Rudkin: page 64.
Steve Owen: page 65.
Bernard Dupont/Flickr: page 66.

Our Non-native Frogs and Toads
Shutterstock: pages 68, 71, 72 (top).
Tim Vickers/Wikimedia: page 72 (bottom).

Frogs Around the World
Shutterstock: pages 74, 77, 78, 79 (top), 80.
Rawpixel free download: page 79 (bottom).
Sathyabhama Das Biju/wikimedia: page 81 (bottom).
Emőke Dénes/wikimedia: page 81 (top).

Threats and Dangers to Frogs
Sean Rudkin: pages 82, 92.
Andy Morffew/Wikimedia: page 85.
Mallaurie Brach/Wikimedia: page 86 (top).
Forrest Brem/Wikimedia: page 86 (bottom).
Shutterstock: pages 89, 91.

John Spiers: page 93.

Under the Knife
Wikimedia: page 94.
Public domain: page 96.
Shutterstock: pages 97, 99.

Frog Spotting
Sean Rudkin: pages 100, 103, 104, 106, 110.
Steve Owen: pages 107, 111.
Tanya Ware: page 108.

Frogs in Myth and Legend
Wikimedia public domain: pages 112, 115, 116, 118, 120.
Shutterstock: pages 121.
Public domain www.lacma.org: page 122.
nativeplanet.com: page 125.

Frogs in Art and Literature
Wikimedia public domain: pages 126, 129, 130, 133, 138, 139.
Project Gutenberg: pages 137, 151-155.
Shutterstock: page 140.
Public domain: page 142.
Kate Wyatt: page 145.
www.stampsoftheworld.co.uk/wiki/
Ireland: page 146.

Jane Russ: linocut endpapers and above.

Every effort has been made to trace copyright holders of material and acknowledge permission for this publication. The publisher apologises for any errors or omissions to rights holders and would be grateful for notification of credits and corrections that should be included in future reprints or editions of this book.

Acknowledgements

A massive thank you to Jane Russ
for inviting me to take part in another
wonderful project and, with frogs and toads
so close to my heart, this has been an utter
joy to be involved in. As ever, working with
Jane is brilliant fun!

A huge thank you to all the talented
photographers and artists who have let us
use their work. Given you're all likely to
have been sitting in mostly damp conditions
for a lot of these images, a doubly huge
'thank you!'

Big thank you to Joana at Graffeg for her
talents and efforts in pulling all the bits
together – it looks beautiful!

To the marvellous Marc Baldwin, who
has, once again, sifted through my words
ensuring I've got the frog's head on the right
way round – thank you, thank you so much
for your time and patience.

My wonderful partner Sean – this is all his
fault, it is! 'Let's look for frogs!' he says, and
ignites a magical passion and curiosity for
the truly remarkable natural world. Thank
you. Love you.

Books in the series

The Hare Book

The Fox Book

The Owl Book

The Red Squirrel Book

The Bee Book

The Robin Book

The Badger Book

The Hedgehog Book

The Native Pony Book

The Puffin Book

The Beaver Book

The Otter Book

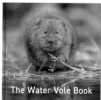

The Water Vole Book

www.graffeg.com

The Frog Book
Published in Great Britain in 2023 by Graffeg
Limited.

Written by Jo Byrne copyright © 2023.
Designed and produced by Graffeg Limited
copyright © 2023. Series editor Jane Russ.

Graffeg Limited, 24 Stradey Park Business
Centre, Mwrwg Road, Llangennech, Llanelli,
Carmarthenshire, SA14 8YP, Wales, UK.
Tel: 01554 824000. www.graffeg.com.

Jo Byrne is hereby identified as the author of
this work in accordance with section 77 of the
Copyright, Designs and Patents Act 1988.

A CIP Catalogue record for this book is
available from the British Library.

Printed in China TT140223

ISBN 9781802583557

1 2 3 4 5 6 7 8 9

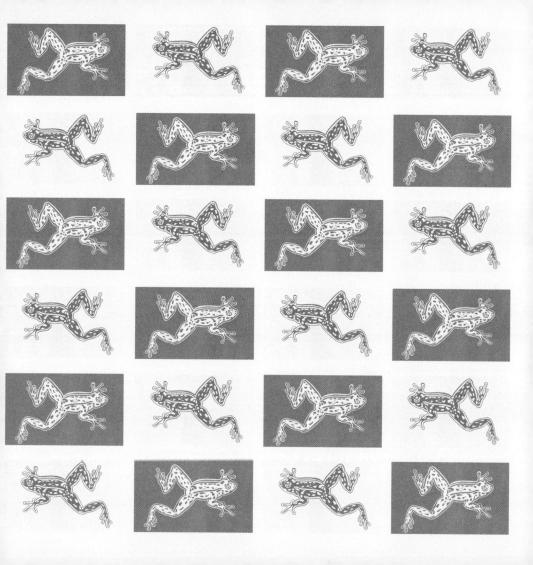

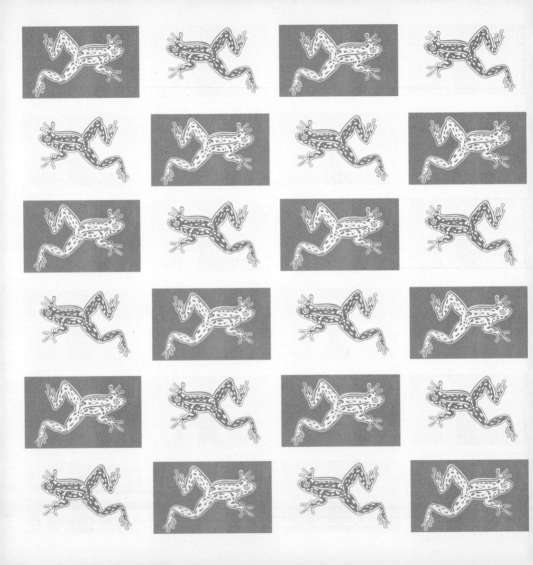